Today is the class trip. Is the class set to go?

"The park is not far," said Miss Clark. "We can start to go!"

“Who can get the basket?”
said Miss Clark.
“I can!” said Oscar.

The class is at the park. But where is Oscar?

Some of them step into the
warm pond.

Some of them get stuff from the art kit.

After that, they sit together. This is the best part of the trip!

"Oscar had a hard job!" said Miss Clark.